I0409436

Abacus an Introduction

The abacus has been in existence for more than 2,300 years. This device was once the primary accounting tool for many forms of business. Several cultures developed their unique style of abacus. Cultures in Asia, India and Middle East still use the abacus to develop mental mathematical skills. Even some merchants continue to rely on the abacus when electric power and mechanical tools are difficult to maintain or not available in a remote area.

One version of a Russian abacus has rods running horizontally and beads of two or more colors strung on the rods.

Счеты
(phonetically:
Schety)

The Japanese abacus may have a rectangular frame with a horizontal bar with one bead above and four beads below the bar, on each of several rods.

そろばん
(Soro ban)

In this tutorial I will use the Chinese Abacus having a rectangular frame and thirteen, more or less, vertical rods or wires, and a horizontal bar dividing the frame into two decks. Each rod, or wire, holds two beads in the top deck and five beads in the bottom deck.

算盘 (Suànpán)

You should use a Chinese Abacus, (Suanpan), with this book. There is no substitute for hands-on practice. Some public Libraries may have an abacus for check out.

An abacus may be purchased for a few dollars.

Some abaci are antiques and many people have extensive collections of them.

It is not a bad idea to learn how to operate more than one style of abacus.

Let the learning begin.

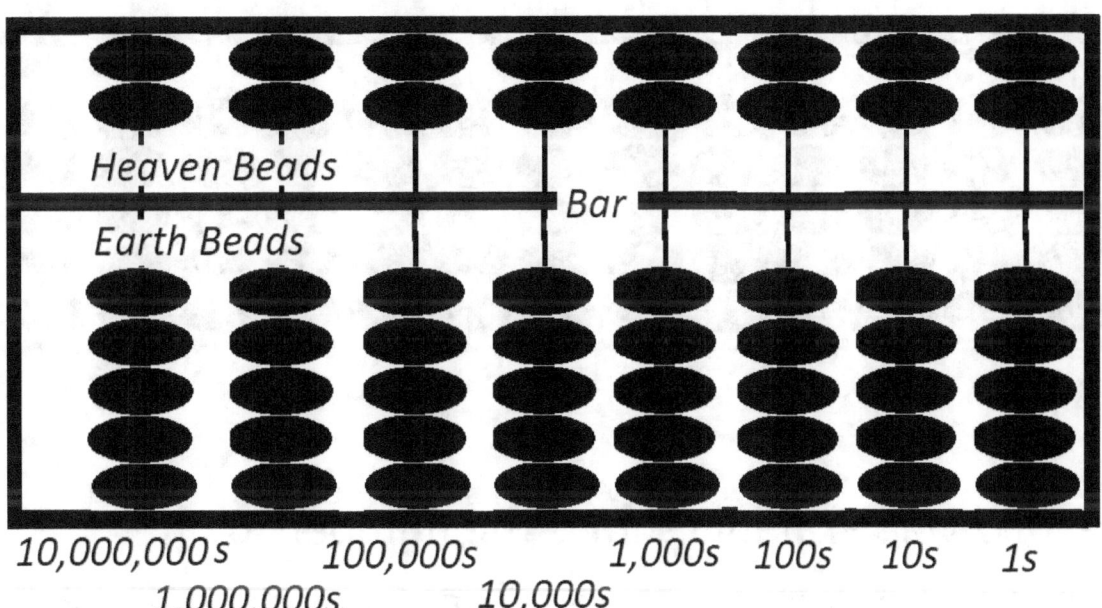

These are the place values for each column

Here is a basic Chinese Abacus.

The Beads in the top deck are worth 5 times their place value. The beads in the bottom deck are worth 1 times their place value. The beads in the top are called "Heaven Beads" and those in the bottom are called "Earth Beads".

A Heaven Bead in a column is worth five Earth Beads in the same column. An Earth Bead in a column to the left is worth two Heaven Beads in the next column to the right.

The following example shows the numbers 6 on column B, 1 on column C, and 5 on column G. Here we are disregarding place value.

An abacus may have any number of columns, colors, or bead values, as long as there is a consistent relationship between beads, columns and place values.

If we read the above as all one number, considering place values, with the most right column representing units, then we have the number 6 10,005.

To place all the beads in a starting position, we lay the frame flat, bottom deck toward us, and then tilt the frame toward us. This brings all the Earth Beads to the bottom, and all the Heaven Beads to the bar. We run a finger along the top of the bar to push all Heaven Beads to the top of the frame, away from the bar.

With no beads touching the bar, the value displayed is zero, or starting position.

Look at the rightmost column. Move the Earth Beads

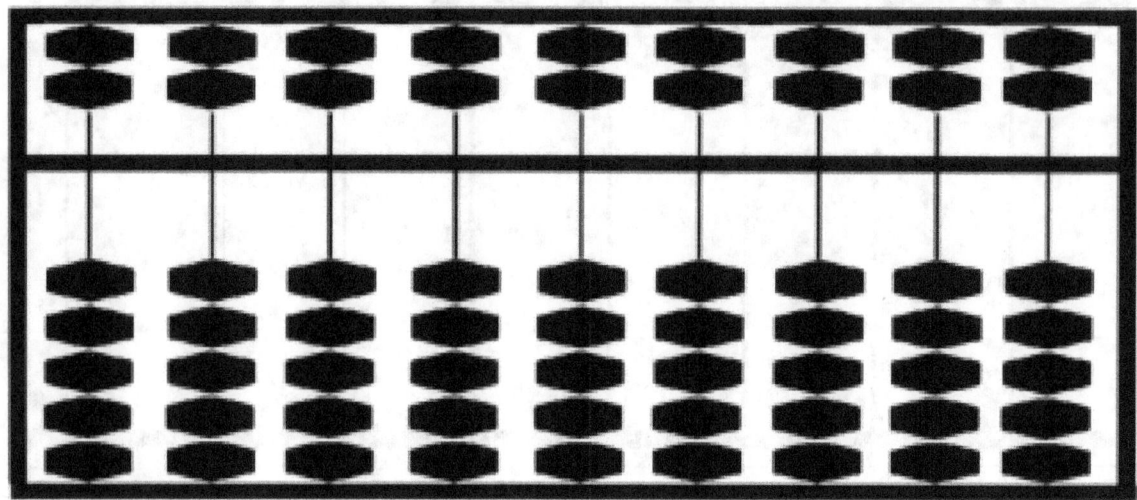

up toward the bar as you count them 1, 2, 3, 4, 5.

The number at the bar is now 5.

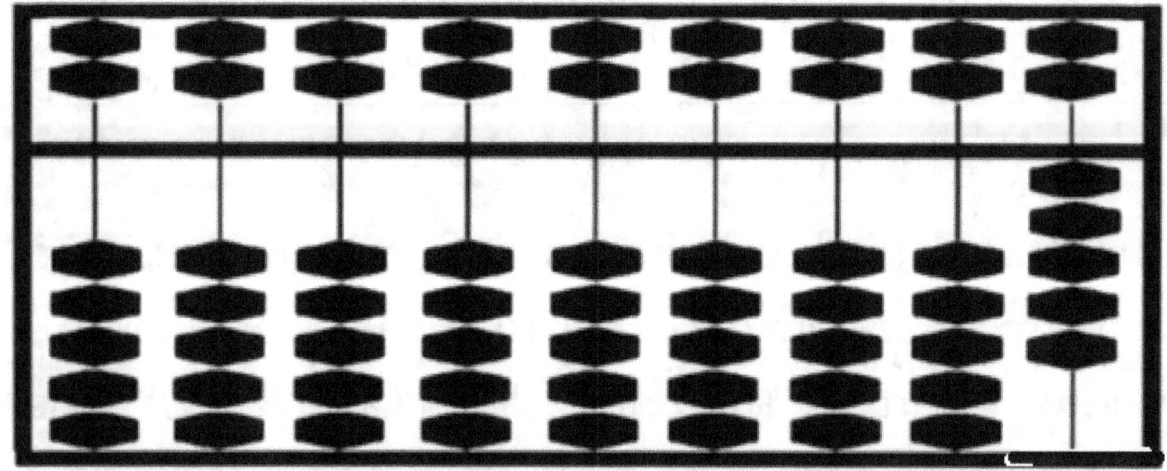

Now, we want to keep counting, so we move 1 heaven bead, on the same column, to the bar; this replaces the 5 Earth Beads which are moved back to the bottom.

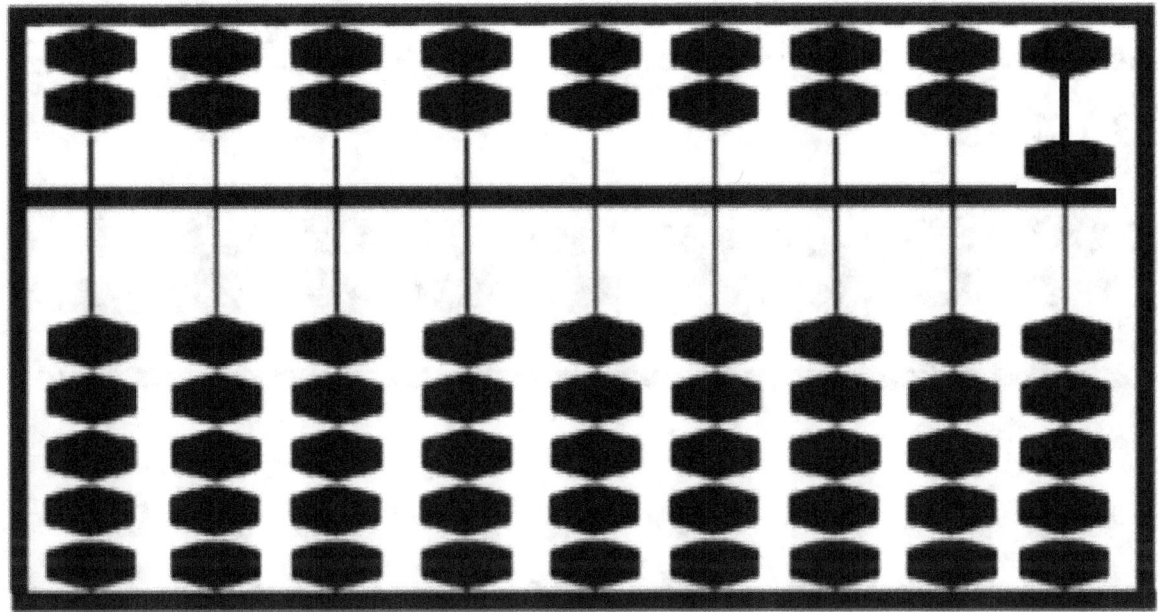

Let us continue to count the same beads, 6, 7, 8, 9, 10.

Again, we are out of units to count, so we move a second Heaven Bead toward the bar on the 1s column, to replace the 5 Earth Beads.

Now we can count the Earth Beads a third time, 11, 12, 13, 14, 15.

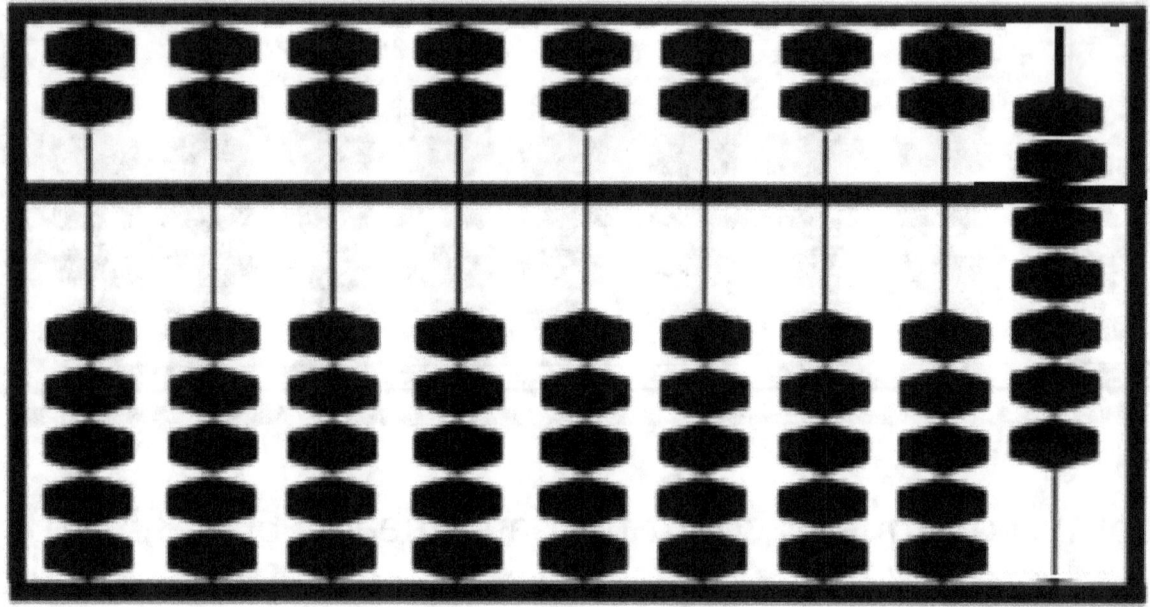

There are 2 Heaven beads (5 each), and 5 Earth Beads (1 each) at the bar for a total number of 15 at the bar.

In order to keep counting we go to the next column, to the left. We move one Earth Bead, on that column, up to the bar. 1 Earth Bead is equal to two Heaven Beads on the next column to the right.

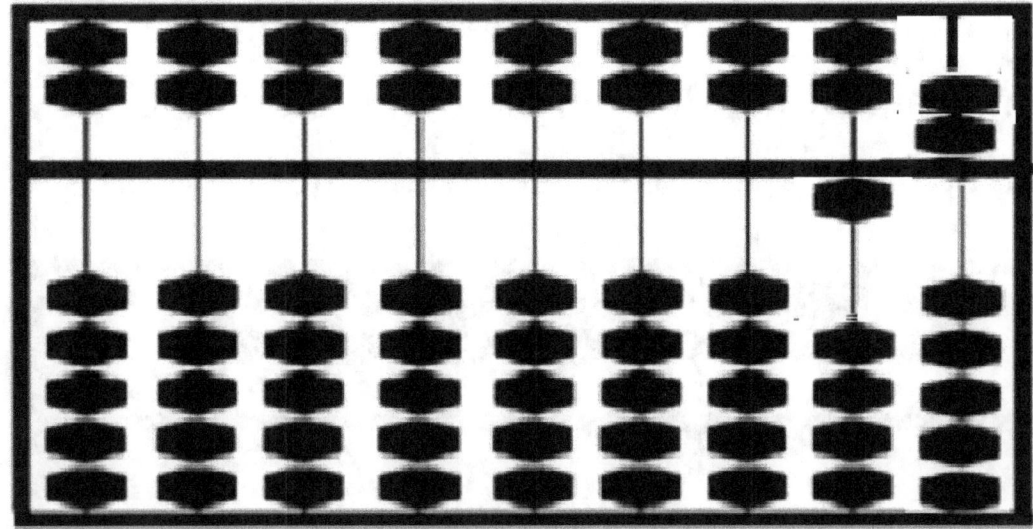

Every time we create a column with two Heaven Beads at the bar, we go to the next column on the left, and replace them with an Earth Bead which is the same value as the two Heaven Beads on the right.

This replaces the two Heaven Beads at the bar! The replaced beads are moved back to starting position away from the bar.

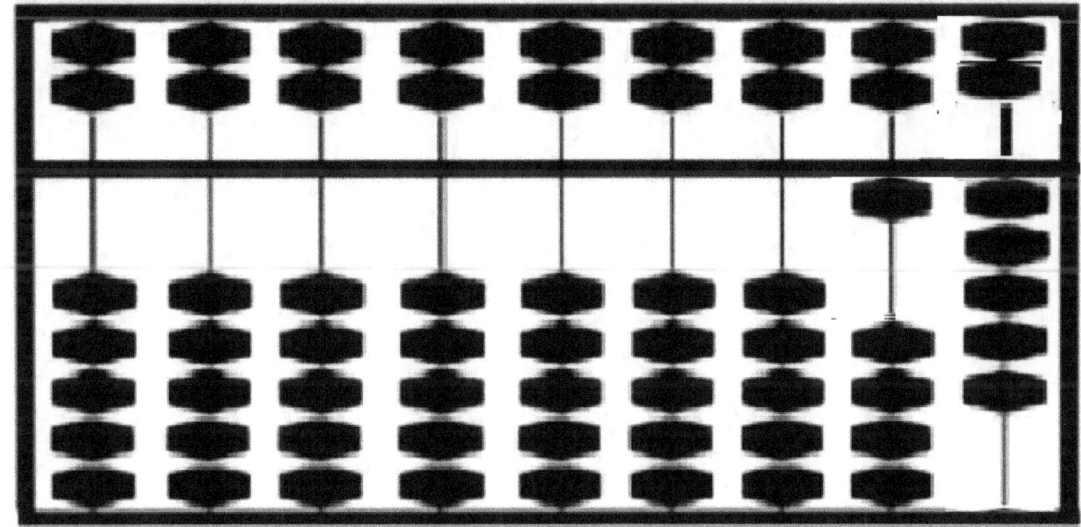

Now we can replace the Earth Beads, on the first column, again with a Heaven Bead and move the five Earth Beads away from the bar on the first column.

The beads touching the bar have the values of a 10 and a 5. There is, still 15 at the bar.

Whenever we have all five Earth Beads up to the bar we need to move a Heaven Bead on the same column down to the bar and move the Earth Beads back to the bottom on that column.

Addition

When we add we simply move beads toward the bar.

Set 13 in columns L and M.

(Red shows which beads are moved.)

We can add to 13 by moving additional beads to the bar. Add 12 to the 13 at the bar.

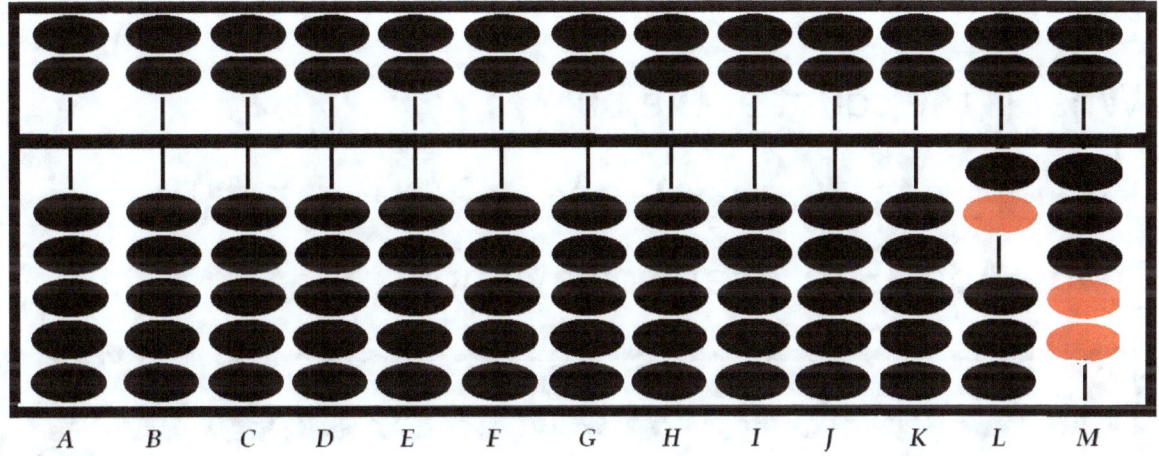

The number at the bar is now 25. Let's add 1000 to 25.

Now we have 1,025 at the bar. When we have put all five Earth Beads in a column up to the bar; we need to reduce that column by replacing the Earth Beads with 1 Heaven Bead from the same column. Reduce column M.

What is the count at the bar?

It is still 1,025; the Heaven Bead replaces 5 Earth Beads. Add 35 to what we have now.

At the bar, is 1,060. Reduce column L.

Whenever we create a column with two Heaven Beads we need to go to the next column left, and move an Earth Bead to the bar. Reduce column M.

The Earth Bead is a replacement for the two Heaven Beads, which are moved away from the bar.

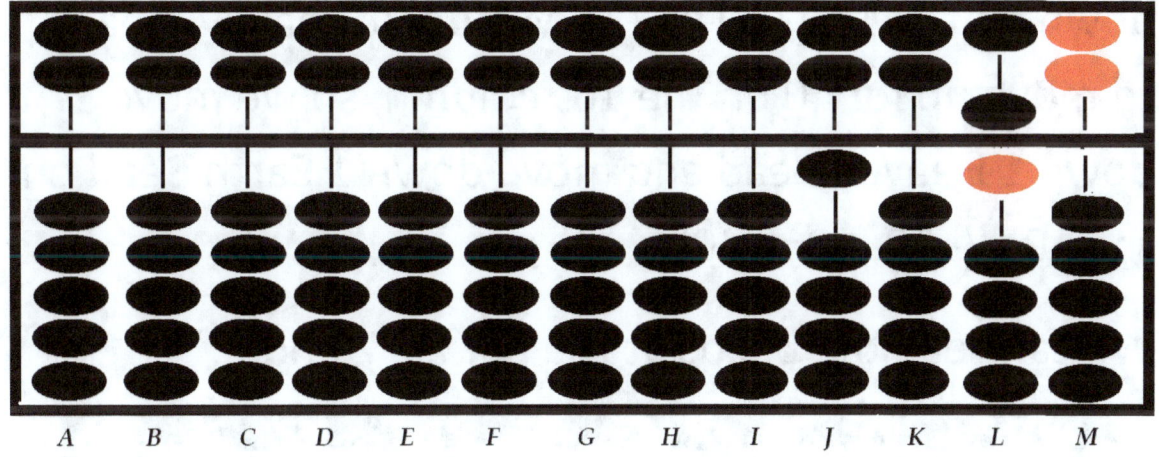

We added 10 on L and removed 10 on M; there was no change in the count at the bar; it is still 1,060!

13

Set 27 in the first two columns on the right. That will be 2 on the 10s column and 7 on the 1s column.

Now we will add 49 to what we have.

We need 1 Heaven Bead and 4 Earth Beads to make 9. We move a bead up on the 10s, and we move a bead down on the 1s. (+10-1=+9). We do not have 4 beads to move up for 40 on the 10s column, so we move down 1 Heaven Bead and move down 1 Earth Bead on the 10s, (+50 -10= +40).

We can read the beads at the bar as, 7 and 6.

(27 + 49 = 76)

Try these on the abacus:

6+8=   12+7=   18+9=   20+14=

175+22=   321+99=   500+108=

Counting & Addition Review

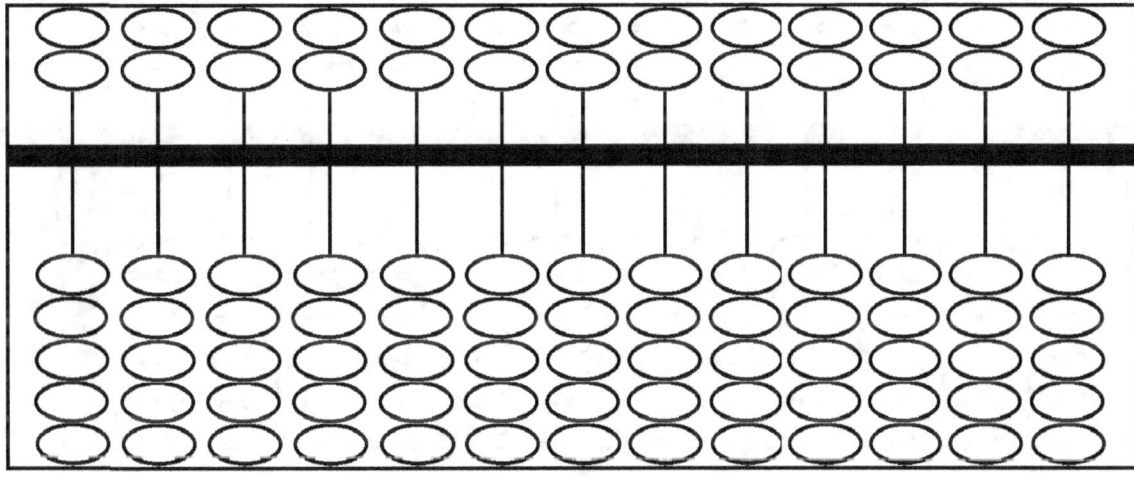

A number is represented by the beads touching the bar, there is no number here.

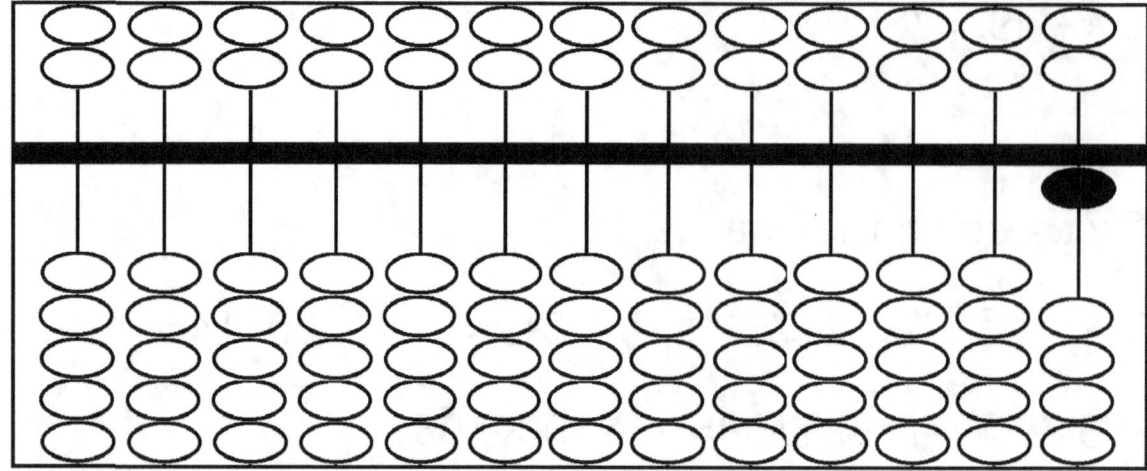

We have moved a bead to represent 1 on the rightmost column.

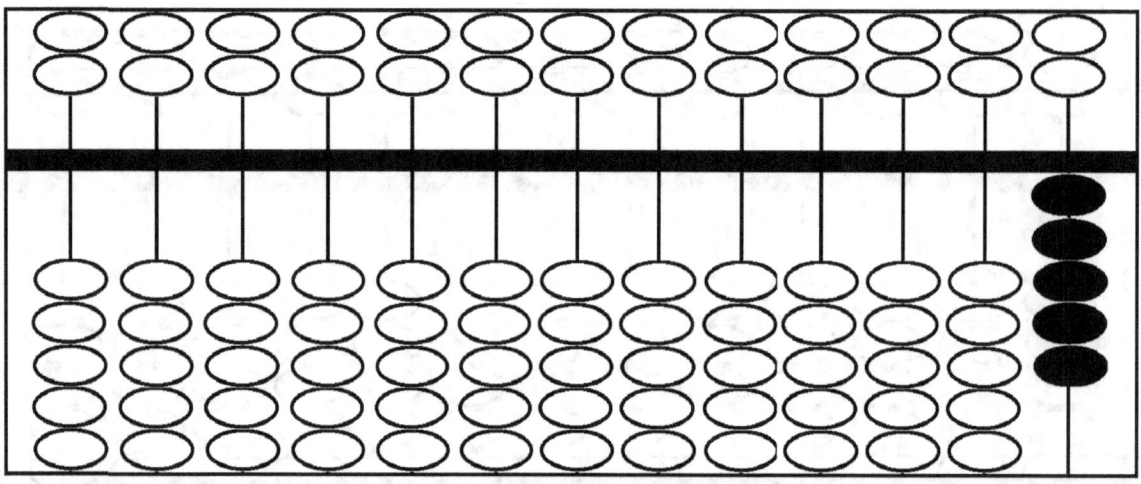

The maximum number of Earth Beads on a column is counted to make 5 at the bar.

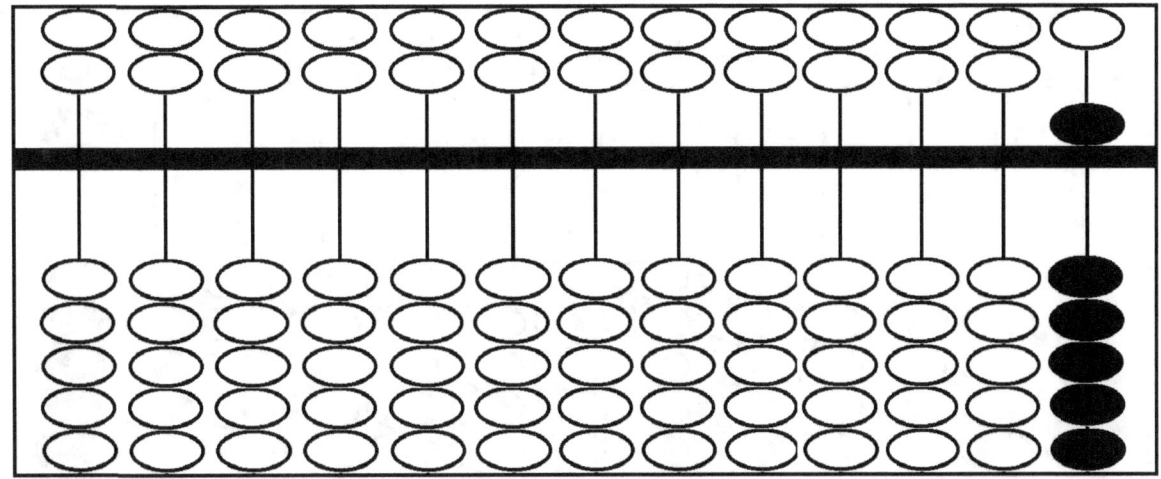

Each Heaven Bead is equal to 5 Earth Beads on the same column, so we can replace the Earth Beads with a Heaven Bead and continue counting with the Earth Beads.

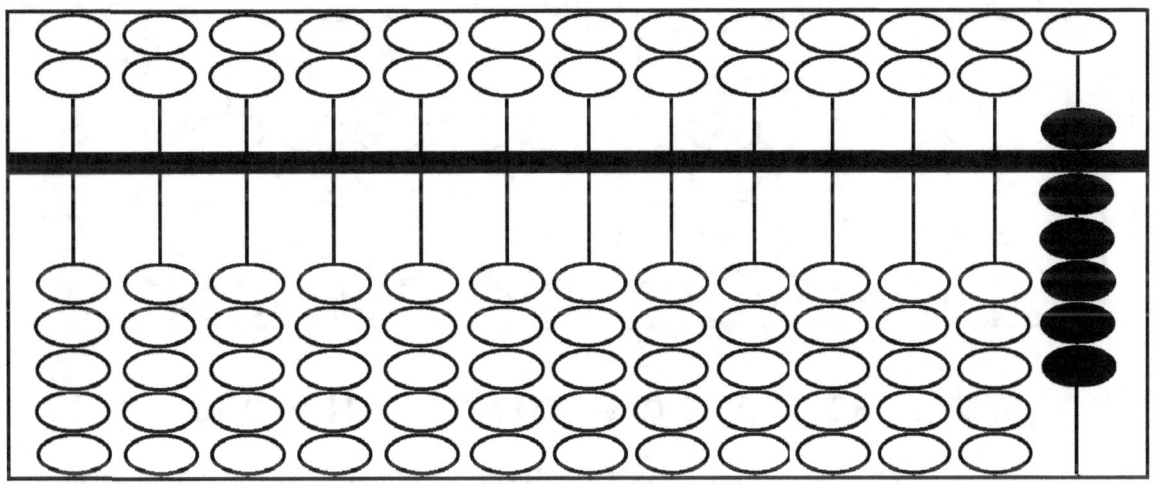

Again, having reached 10 at the bar, we replace the Earth Beads.

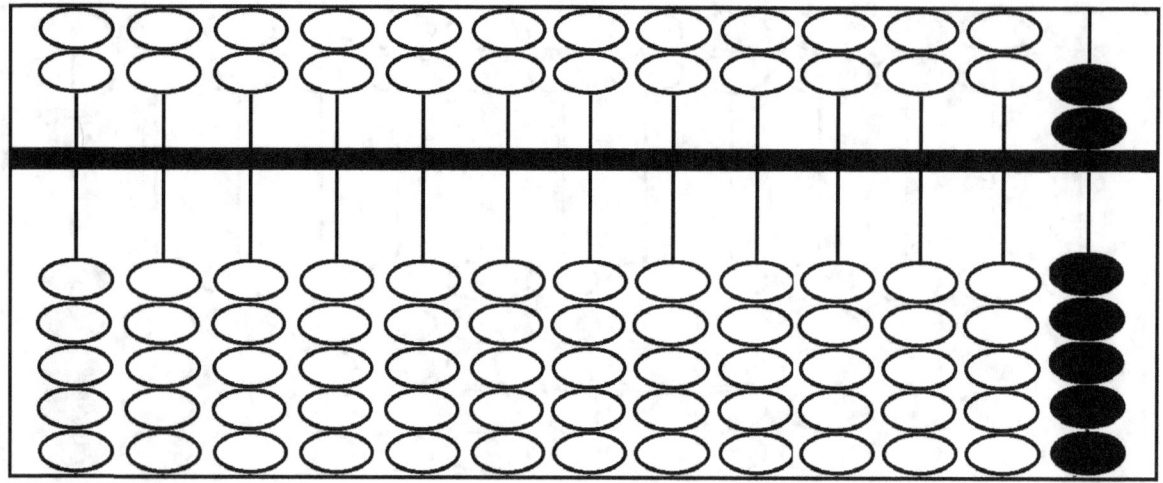

We may use the same Earth Beads to continue a third time. The count is 10 at the bar.

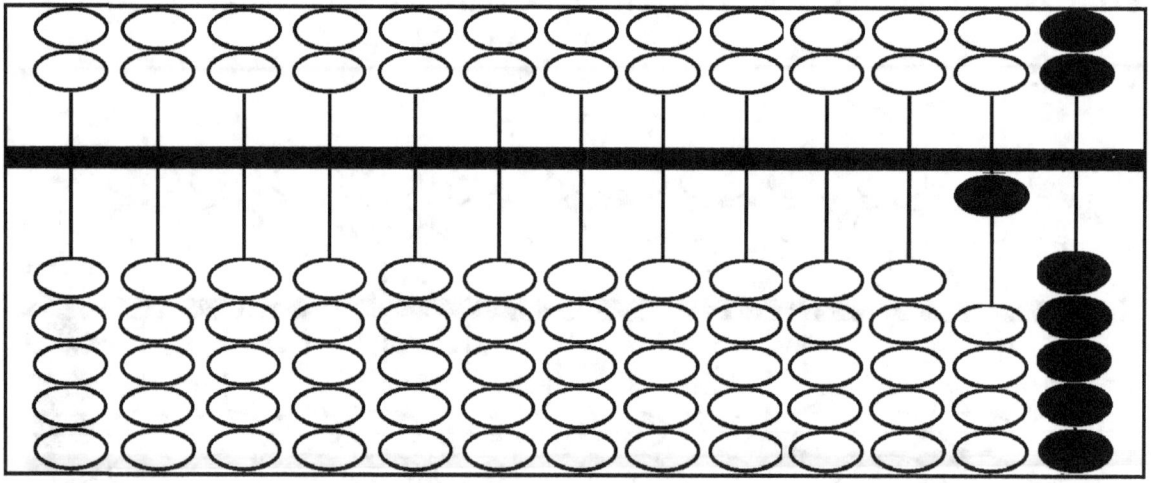

One Earth Bead on the next column left is equal to two Heaven Beads on the next column right. We replace the two Heaven Beads with one bead on the next column. The count is still 10 at the bar.

For Addition, we simply add the beads needed for the number being added to the beads of the current number.

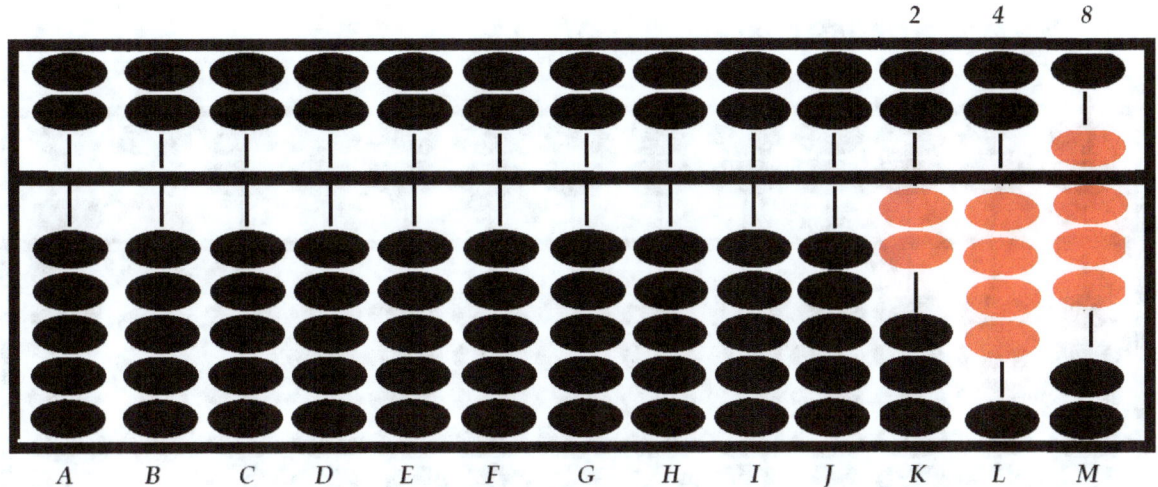

If we need to add units that are not available on the column we need to place them on, then we add a greater number and subtract the excess!

We have 248 at the bar and we want to add 29.

M has only 2 units available so, to add the 9, of 29, to M, we add 10 at L and remove 1 at M.

(+10 -1 = +9) Then we need to add 2 to column L.

L has no Earth Beads away from the bar, so we must first reduce L then add the 2 to L, or a faster way is to add 5 to L and remove 3 from L.

The beads now read 277, (248 + 29).

With 277 on your abacus, add 954.

Reduce

step 2

step 3

The sum is 1,231 and it should look like this.

Try these:

439 + 197 =   90 + 139 =   208 + 780 =   605 + 82 =

Subtraction

To subtract we move beads away from the bar.

Sometimes when subtracting a number we need to do the opposite replacement we do when adding. We may need to subtract a larger number than we intend and add back the excess!

Set 388 on the abacus

A   B   C   D   E   F

Subtract 159

First we can move an Earth Bead down on column D (-100), then on column E we can move a Heaven Bead up (-50), but to subtract the 9, we move an Earth Bead down on column E (-10) and move an Earth bead up on column F (+1).

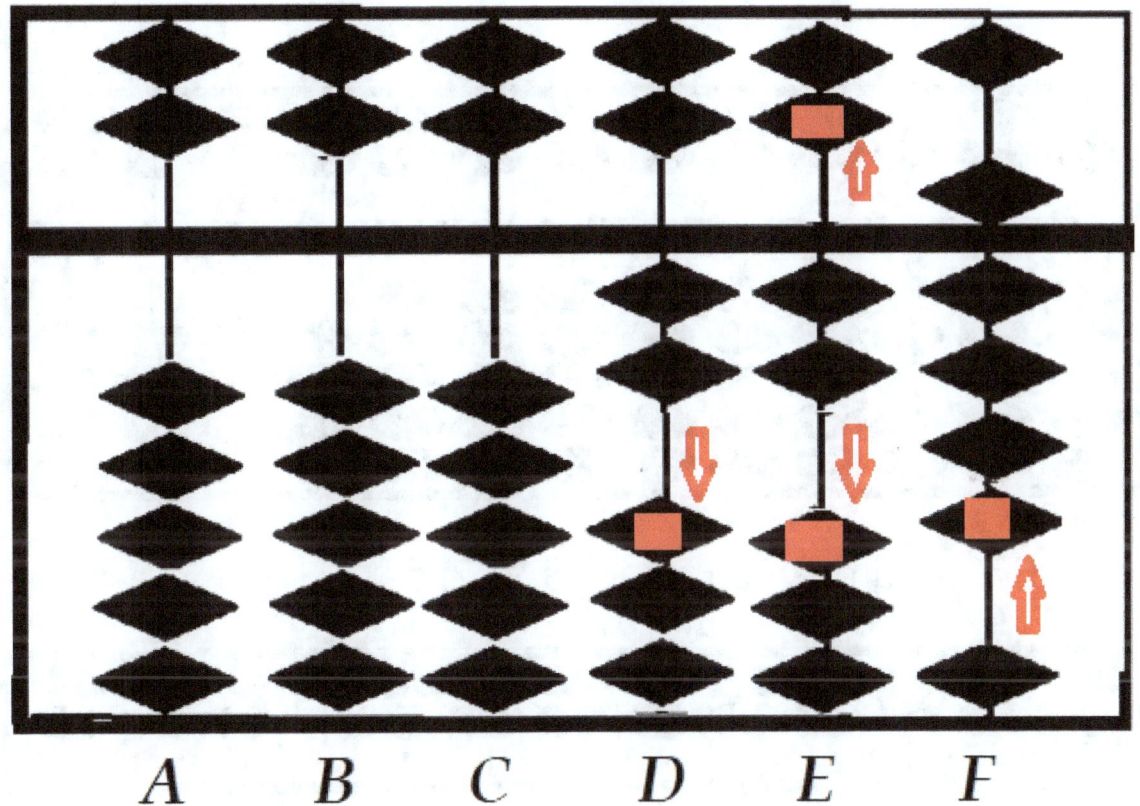

The result is 388-159 = 229.

Set 1,406 then subtract 795.

Set 5,070 and then subtract 3,982.

5,070 − 3,982 = 1,088

If you got this one, then you know how to add and subtract on the abacus.

Try these exercises:

934 − 29 =   1,845 − 590 =   107 − 99 =   919 − 75 =

Multiplication

When we multiply or divide, we will have two numbers on the abacus to carry out our operations.

We want to be sure to leave at least one null column between those numbers and more if we have room, for example: multiplicand and multiplier, or divisor and dividend.

There is always a need for extra caution when working with numbers ending in zero. A column holding a 0 can be mistaken for an empty column. For numbers ending in zero, leave an extra unused column.

On the Chinese abacus the beads are usually all the same color, but in diagrams we try to make it easier to see which beads are moved and which are stationary. Different colors and shadings may be used for this purpose. Unused or uncounted beads may be left out of diagrams altogether. As we multiply we add numbers to the product from right to left, so start your answer as far right on the abacus as you can.

1 digit multiplier

104 x 4

Set 104 to the left and leave at least one column as a separator between the multiplicand and the multiplier.

1   0   4   x   4   A   B   C

104 x 4 = 416

The answer starts with the base for the 4 at column C and it is 16 (4x4). For the next digit of the multiplicand, the base is at column B, and the product is 0.

(4 x 0), the base then moves to A for the next digit and is 4 (4 x 1).

We started the answer at C because we did not know how many spaces (columns), we would need for the answer.

Try these:

15 x 5 =   204 x 6 =   19 x 7 =   129 x 9 =

2 digit multiplier

For the rightmost digit of the multiplier the base is at

EXB = I, EXA = H, DXB = H, DXA = G

the rightmost column.

As the digits of the multiplier are applied to the digits of the multiplicand, the starting base is moved one column to the left for the multiplier.

The base is also moved one place to the left when we use the next digit of the multiplicand, going right to left, in both the multiplier and multiplicand.

Set up 213 x 18

8 x 3 = 24

8 x1 = 8. We add 8 to what we have

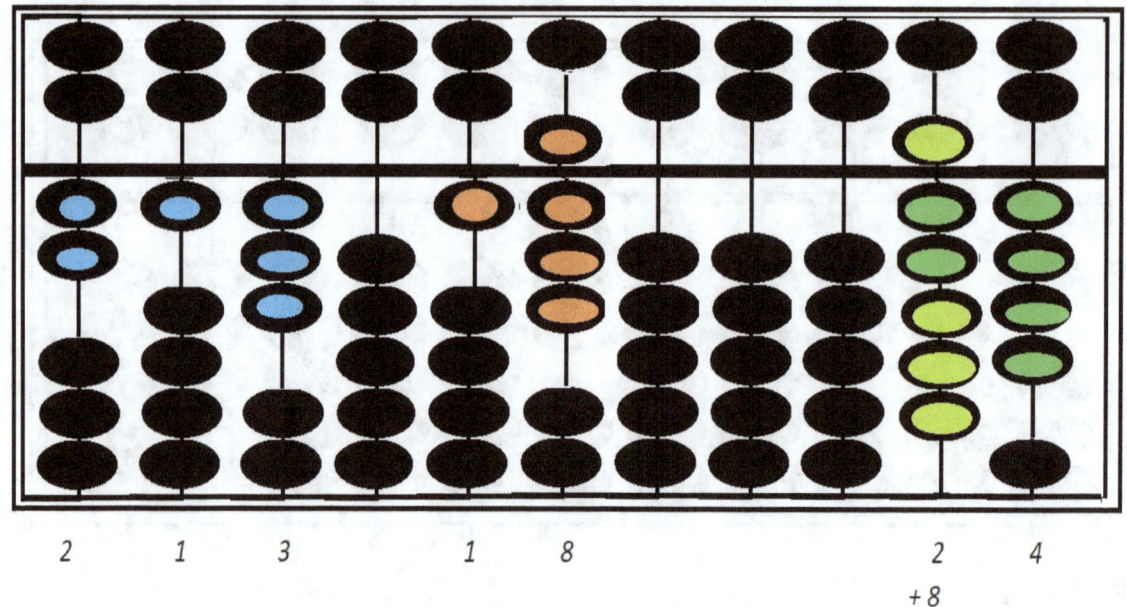

8 x 2 = 16 Added to what we have

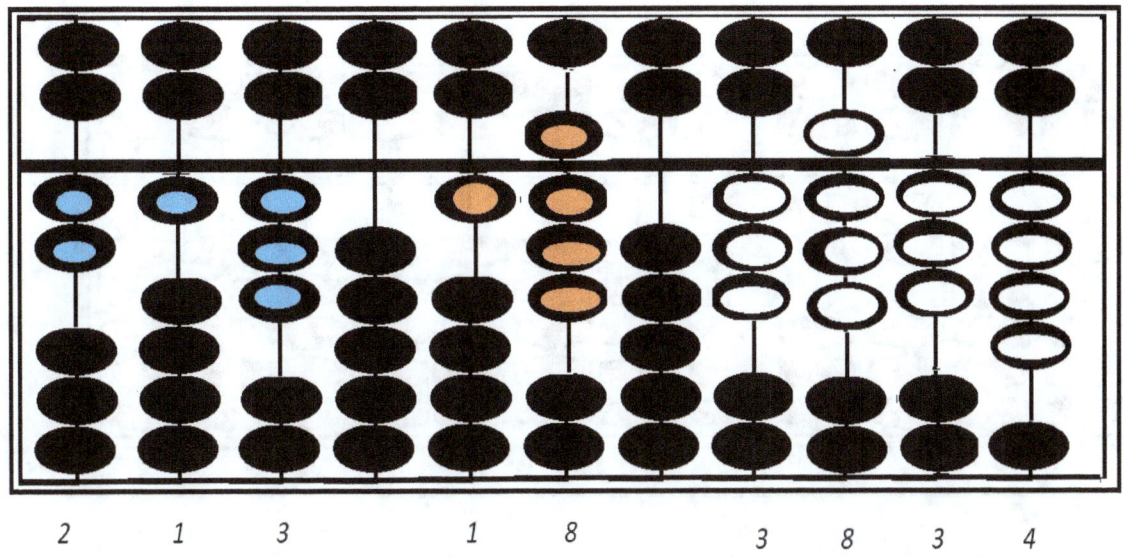

2   1   3   1   8   3   8   3   4

Now the multiplier is the 1 in 18 so the base is at the second column from the right.

1 x 213 = 213, so the solution, when this product is added, is 3,834.

During the calculations we did not forget to adjust when 2 Heaven Beads where at the bar! We did not forget to move an Earth Bead up on the left column when we pushed 2 Heaven Beads on the next right column back to the top! We did not forget to move a Heaven Bead down, on the same column, when we moved 5 Earth Beads down If we did then maybe we have the wrong answer.

3 digit multiplier

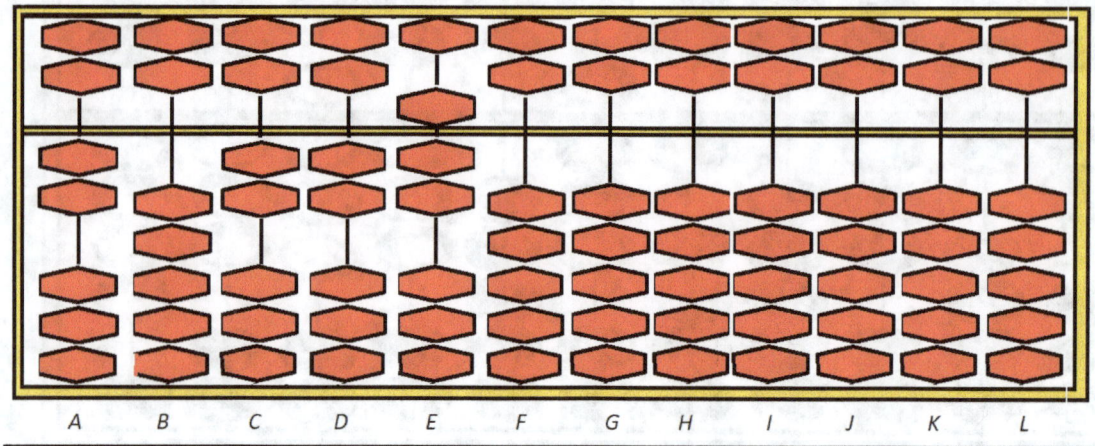

Let us multiply 2 x 227. The base for multiplier 7 on E is at L, for 2 at D the base is at K, and for 2 at C it is J.

7 x 2 = 14 and we set 14 on the right two columns as 1 and 4

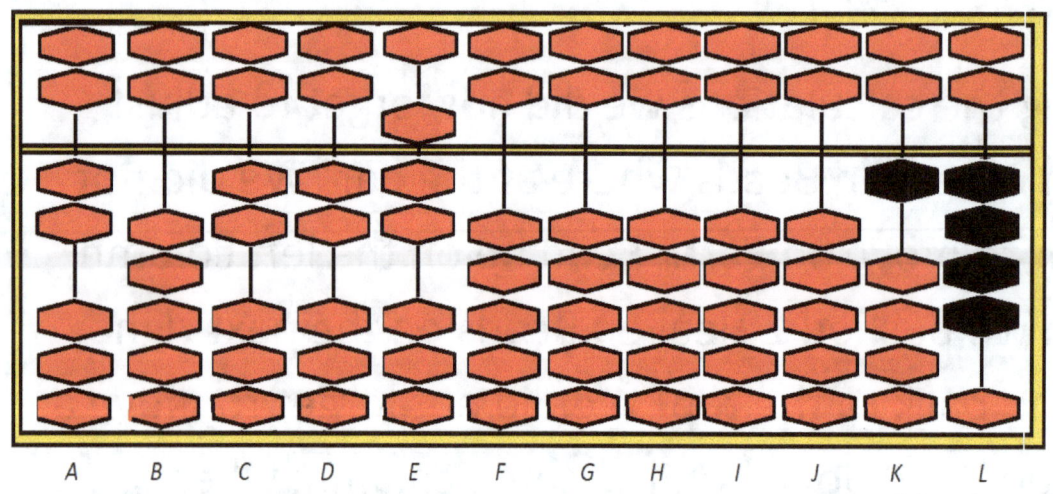

2 x 2 = 4 so we move up 4 Earth Beads on K

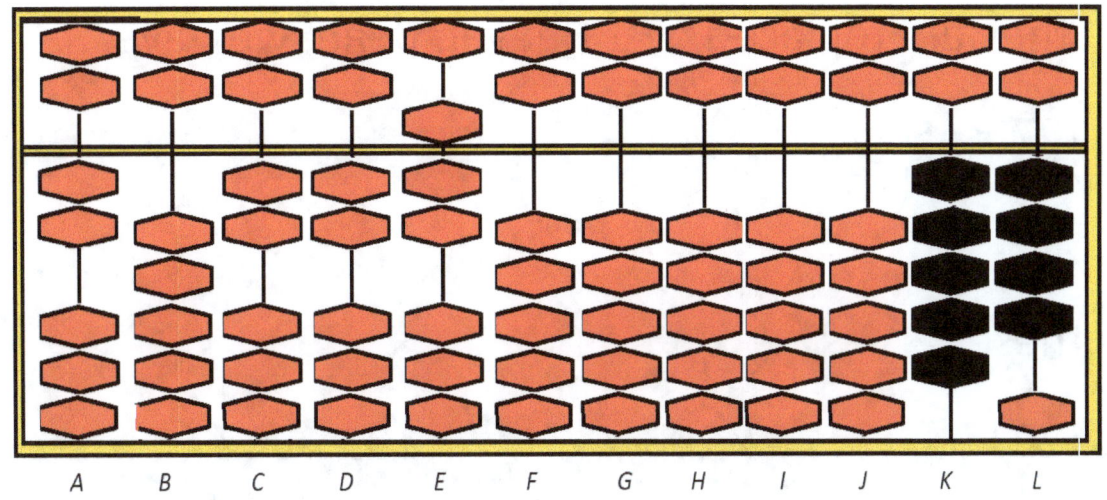

2 x 2 = 4 so now we add 4 Earth Beads on J

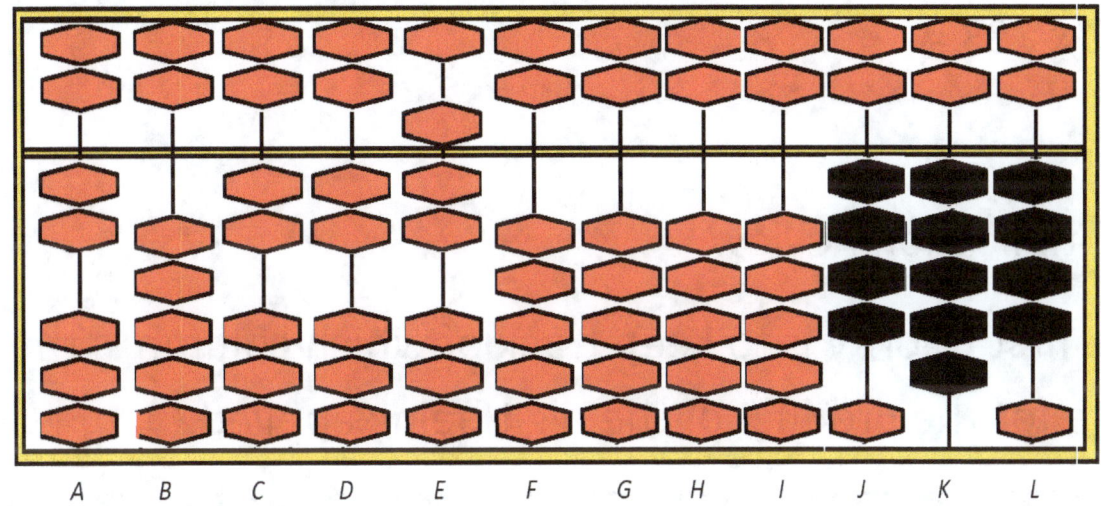

Now we need to adjust for K, all Earth Beads are used

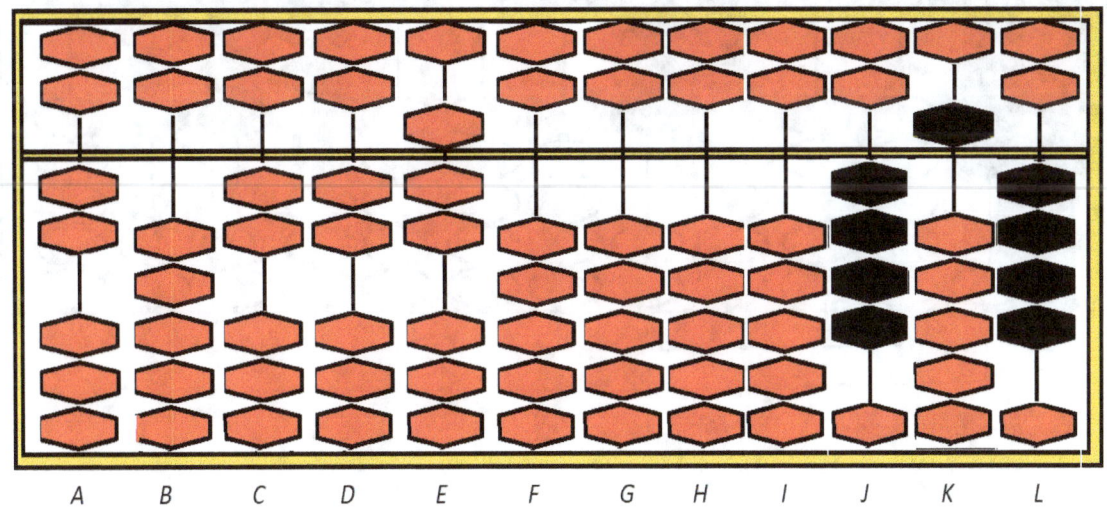

and they are replaced by 1 Heaven Bead, we have

2 x 227 = 454.

Try these..

4 x 129 =   6 x 206 =   109 x 342 =

240 x 120 =   22 x 123 =   118 x 246 =

992 x 183 =

Multiplication Review

The first priority is to keep track of which column is the base for the product of the digit being applied.

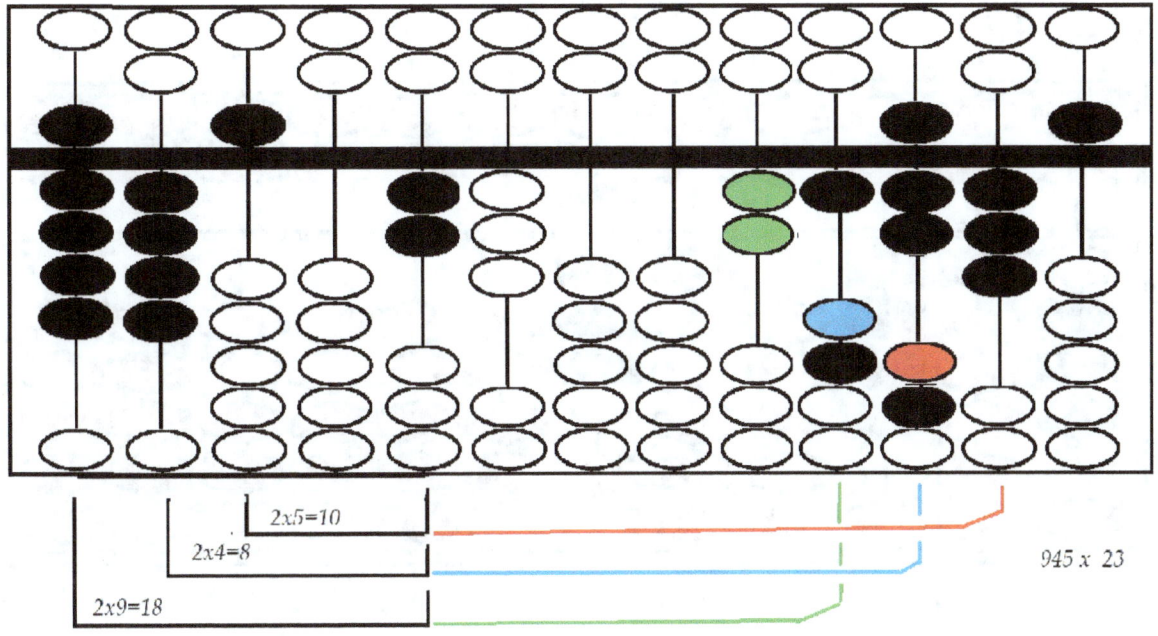

945 x 23 = 21,735

Division

When dividing on the abacus we need to allow sufficient spaces for the quotient as we enter it from left to right, so we start our answer at the leftmost column when possible.

414 / 18

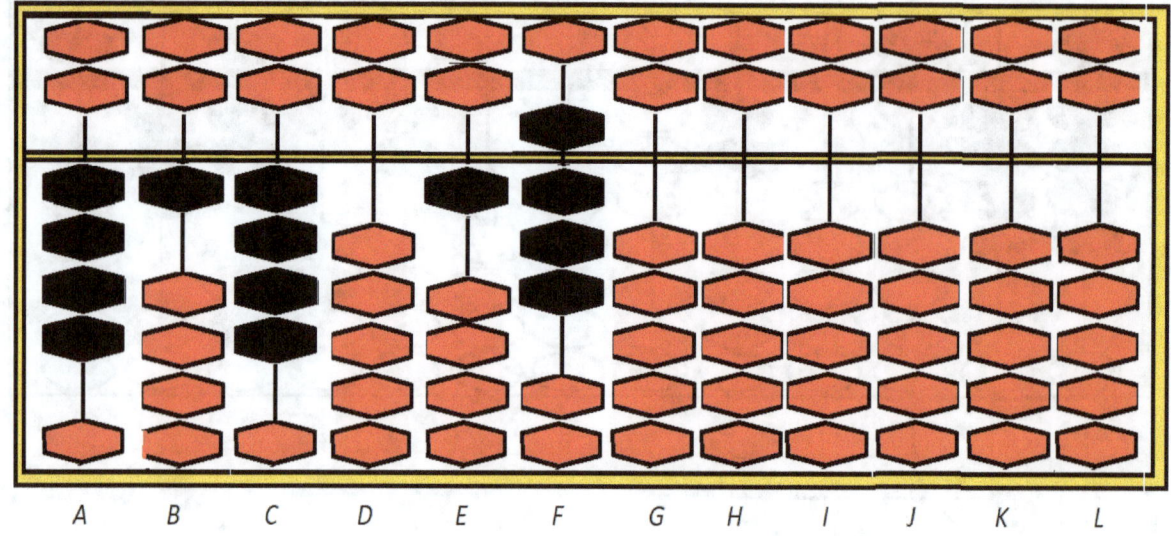

18 goes into 41 twice (2 x 18 = 36), 41 − 36 = 5 (at B)

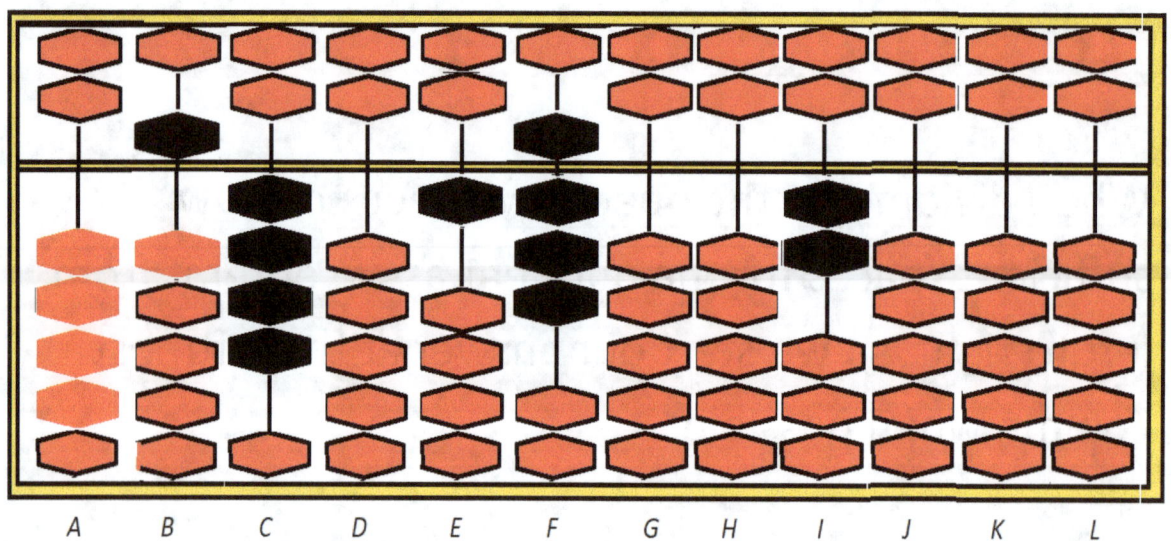

We have reduced the 41 by 36 as you can see on the abacus we now have 54 to divide by 18. 18 goes into 54 three times (3 x 18 = 54).

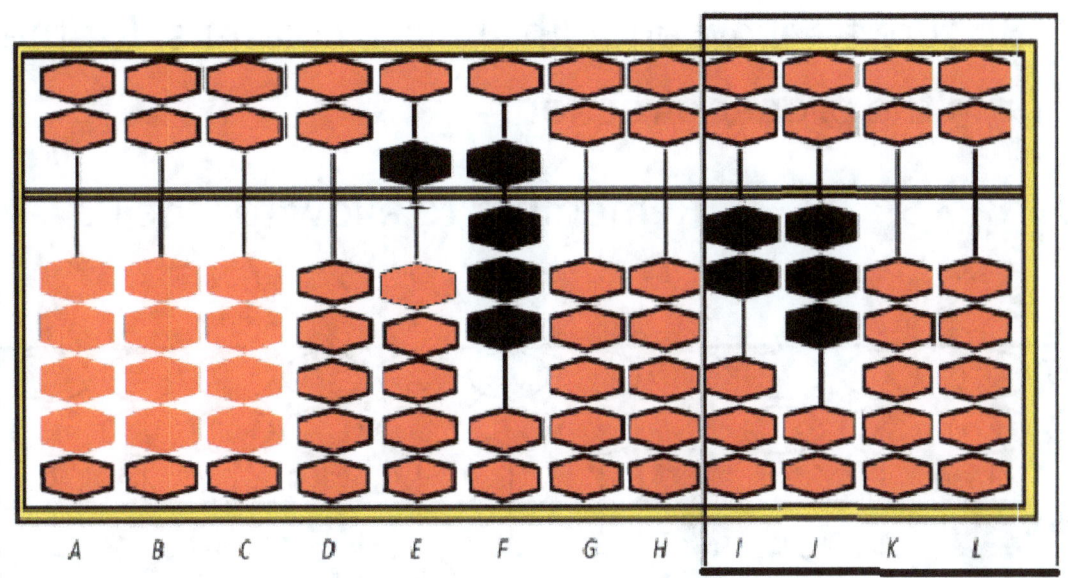

Add the 3 Earth Beads to J. Subtracting 54 leaves us no remainder with 23 as the quotient. We started our answer at column I, leaving 2 decimal places for a possible remainder.

We can also find the quotient by repeatedly subtracting 18 from 414. We would do it 23 times.

Can you complete these?

225 / 25 =   216 / 3 =   608 / 19 =

135 / 15 =   658 / 72 =

Division Review

For multiplication we enter the product from right to left, but for division we enter the quotient from left to right. Be sure to allow enough spaces, columns, for the answer in the proper direction.

(140 x 8) / 5   Remember to allow for the 0.

140 x 8

(1,120) / 5

224

Decimals

On the abacus the decimal point is wherever you designate. In the previous calculation we might have left some columns on the right for any remainder we might have. On the abacus we do not record fractions in any form but decimals. When there is a fraction in a calculation or result, we convert it to the decimal form. The number of decimal places you will need is determined by how accurate your number must be. Most domestic calculations can do with two decimal places.

You may have more than one number with decimal points on the abacus at the same time.

32.45 + 3.1416 (keep all decimal points lined up vertically when adding and subtracting, allowing space for the one with the most decimal places.)

$$
\begin{array}{r}
32.4500 \\
+ \quad 03.1416 \\
\hline
35.5916
\end{array}
$$

We need to plan for at least 4 decimal places!

Set 32.4500

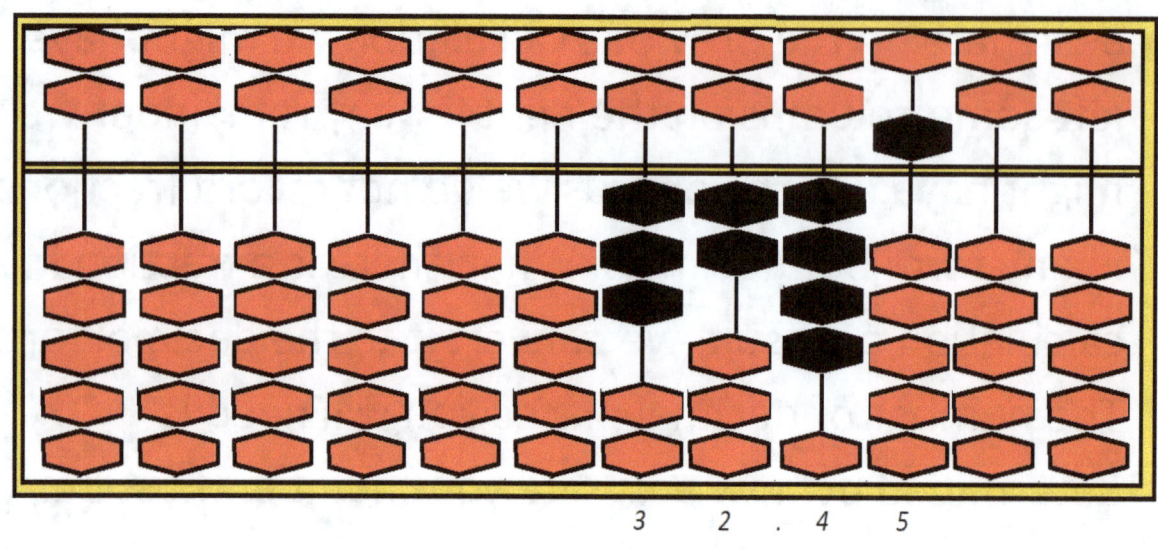

3   2   .   4   5

+ 3.1416

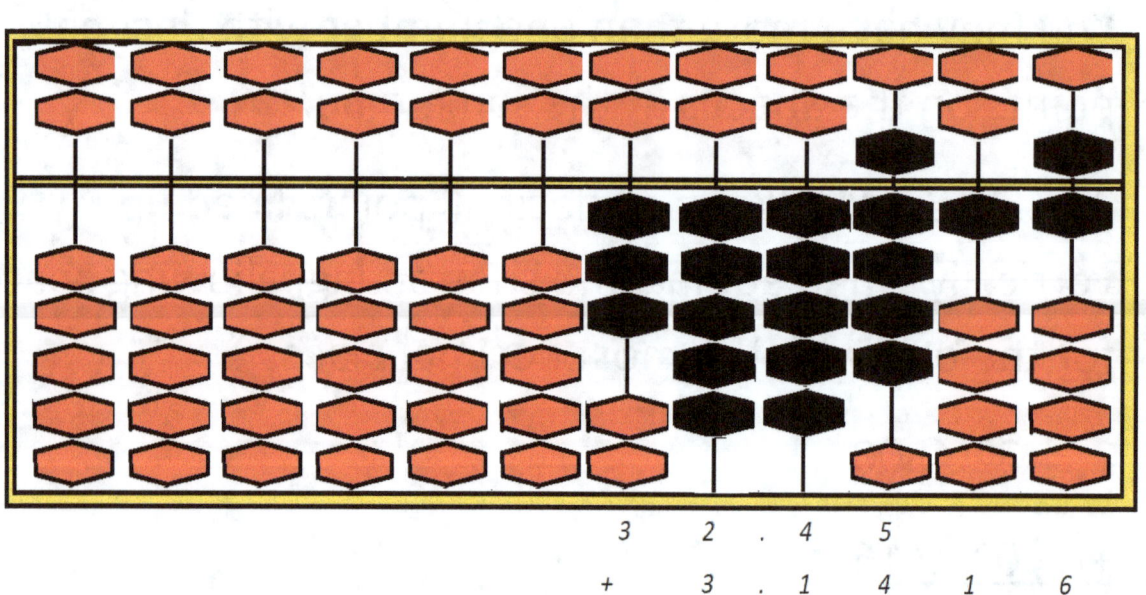

3   2   .   4   5

\+   3   .   1   4   1   6

Replacing 5 Earth Beads with 1 Heaven Bead: then

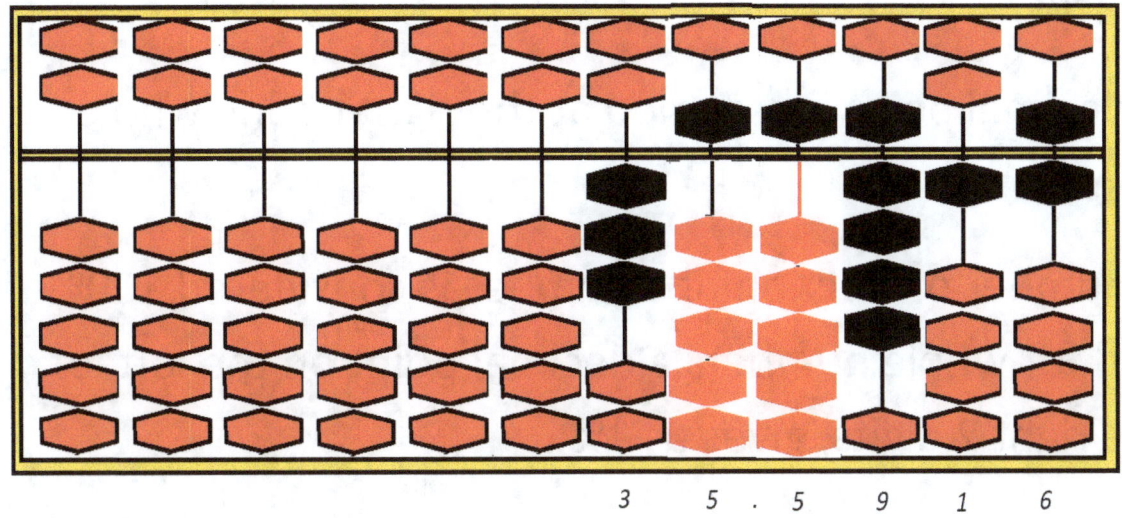

3   5 . 5   9   1   6

read 35.5916

Percents

Percent means "for each 100".

Most financial calculations referring to US dollars require no more than 2 decimal places. There are some values in dealing with stocks and bonds trading that require three or four decimal places.

When a calculation result contains a fraction or a remainder, we need to convert it to the nearest decimal equivalent in order to work with it on the abacus.

A decimal point is assumed at the right of any whole number.

A number or fraction of a number is considered 100% and for whole numbers a decimal point on the left, of the right 2 digits, gives us 1%.

Fractions are converted by dividing 100 by the denominator and multiplying the numerator by that factor.

2/25, 100/25 = 4, 4 x 2 = 8: 2/25 = 8% or .08

What is 5/20 x 6/24? .25 x .25 = .0625 (total decimal places in factors is 4 so we allow for 4 places in the answer)

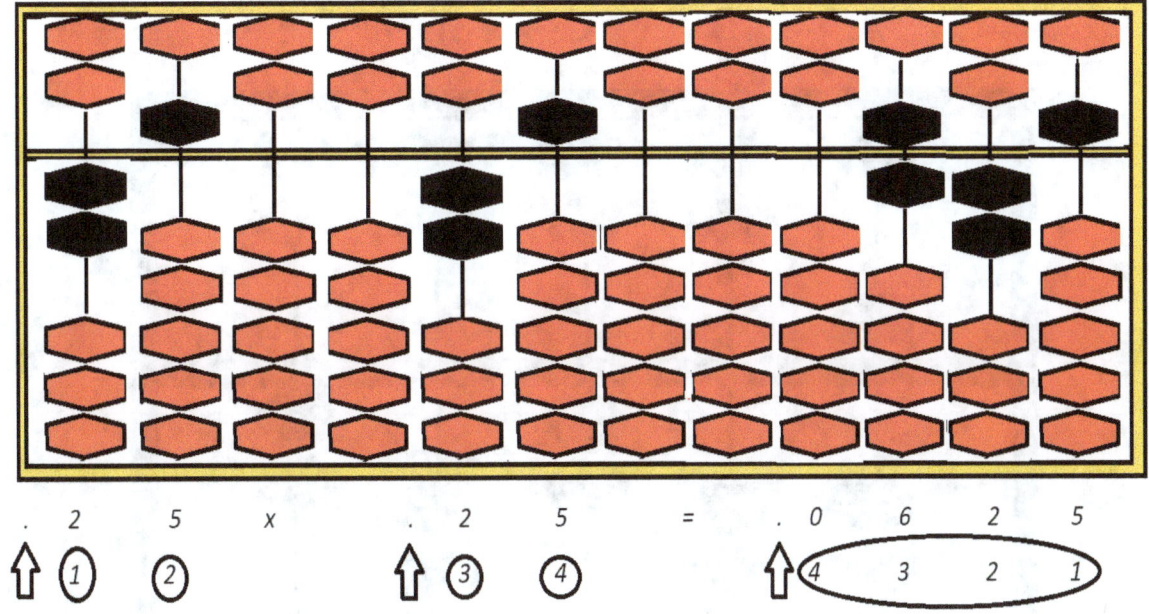

Please total these:

100-23 + 840- 19 -8 + 223- 15 +230 -102 =

(12/3 + 22) x ((6 + 14) x 15))-195 + (234/ 468) =

There are at this time several Abacus tests and certifications available online or by mail. There is no ultimate authority for certifying Abacus Expertise in the United States. A certificate from one school or instructor is as good as any other.

You will have earned your certificate for this course by successful completion.

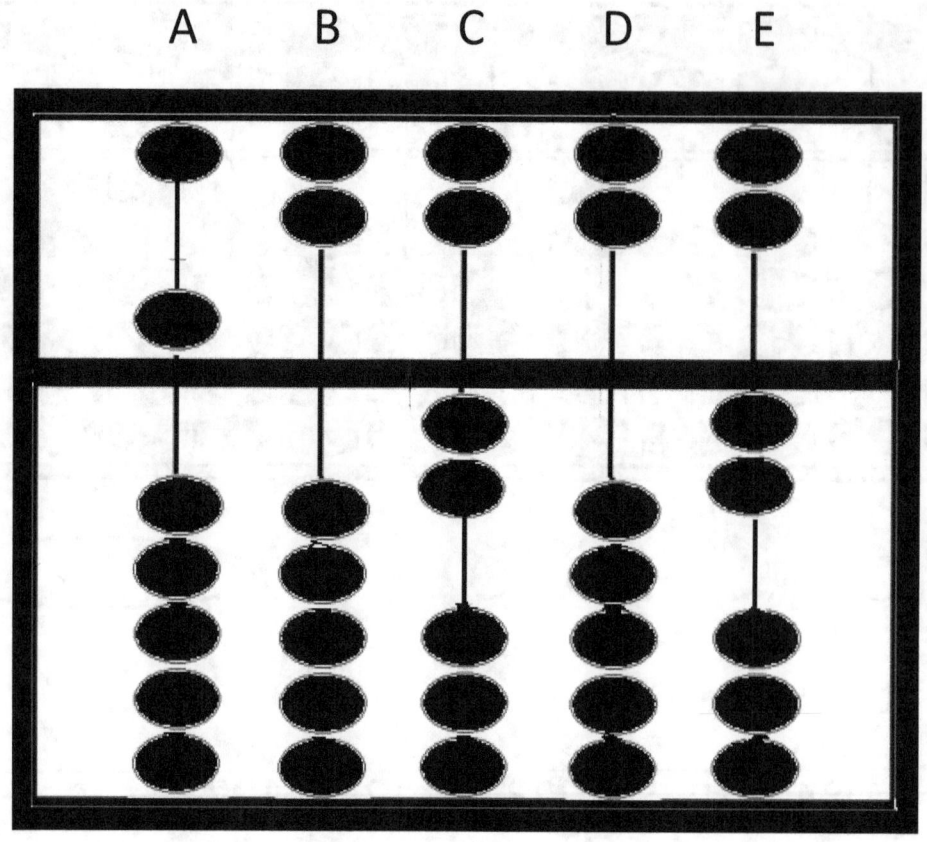

The Decimal point is between C and D

What do you read this number to be?

10202_____   502.02_____   5.202_____   .5202_____

a   b   What do we do next?

Move all beads on a up_____

Move a Heaven Bead down and
all Earth Beads up on a _____

Move down all beads on b____

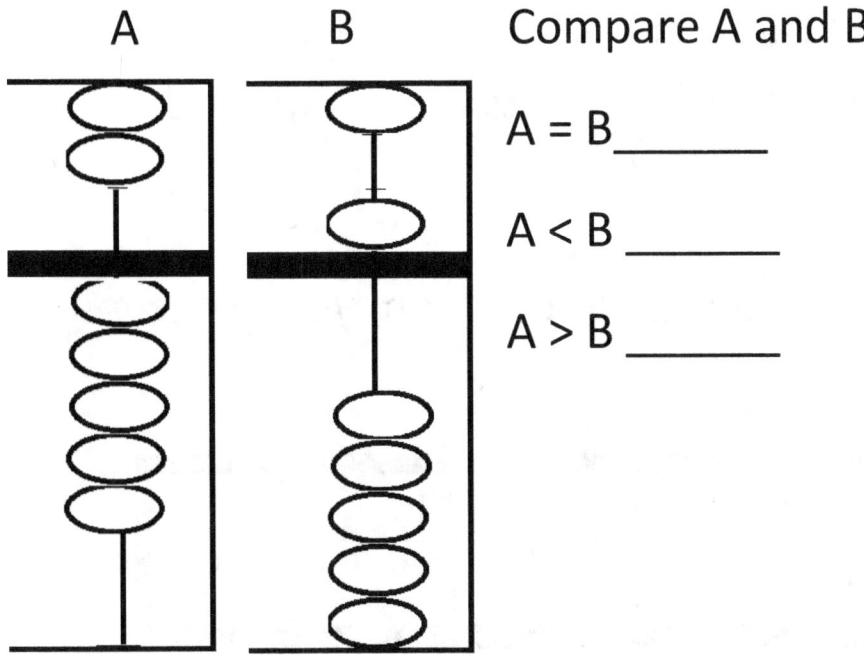

Compare A and B

A = B _____

A < B _____

A > B _____

Which comes first and gets adjusted to the other?

C then D _____

Neither _____

D then C _____

C then D then D _____

The following is a chance for you to test yourself and determine your level of Chinese Abacus ability.

Answers will be at the end of this book and should not be read until all parts of the tests are completed.

Test A

Draw the answer beads positions from your abacus.

a. 22 + 34

b. 16+26

c. 104+220

d. 30+70

e. 4004+2030

f. 180+810

g. 999+499

h. 115+15

i. 203+219

j. 4009+9090

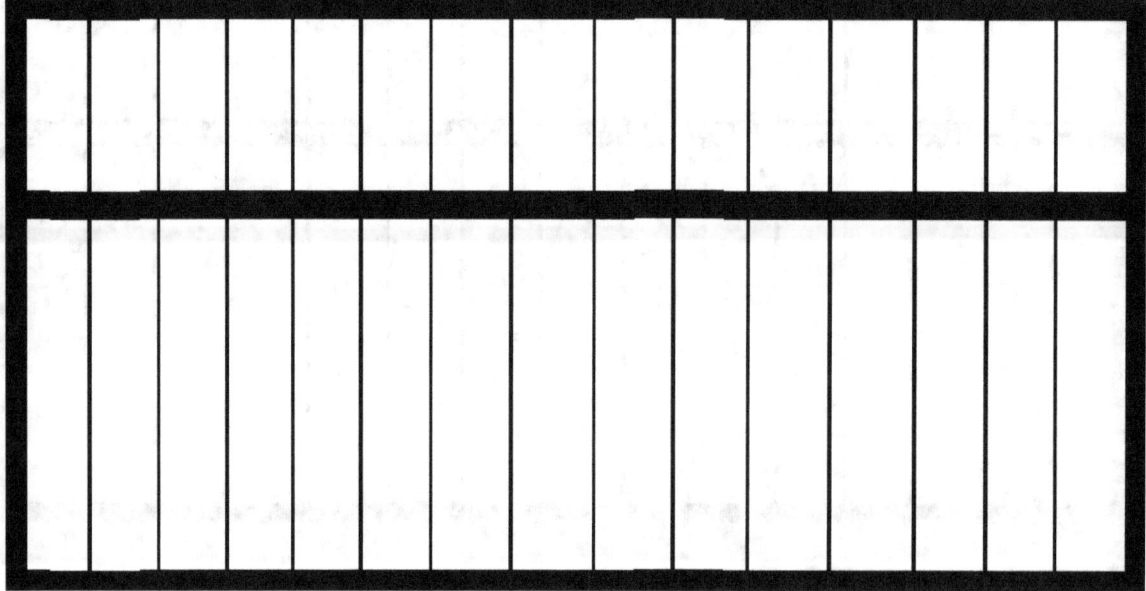

Test B, Diagram beads of result from your abacus

a. 35-18

b. 355-28

c. 298-120

d. 507-203

e. 3001-2003

f. 100-97

g. 222-199

h. 606-284

i. 903-605

j. 505-209

Test C, Diagram solution calculated on your abacus

a. 23+18-9+17

b. 304-210+190-12

c. 213-31+180

d. 108-9+19-26

e. 999-18+54-22+75

f. 211+112-121+212-2

g. 309-27+91-100

h. 209-87+91-17+20

i. 120-19+109-15-22+390

j. 3400+602-1009

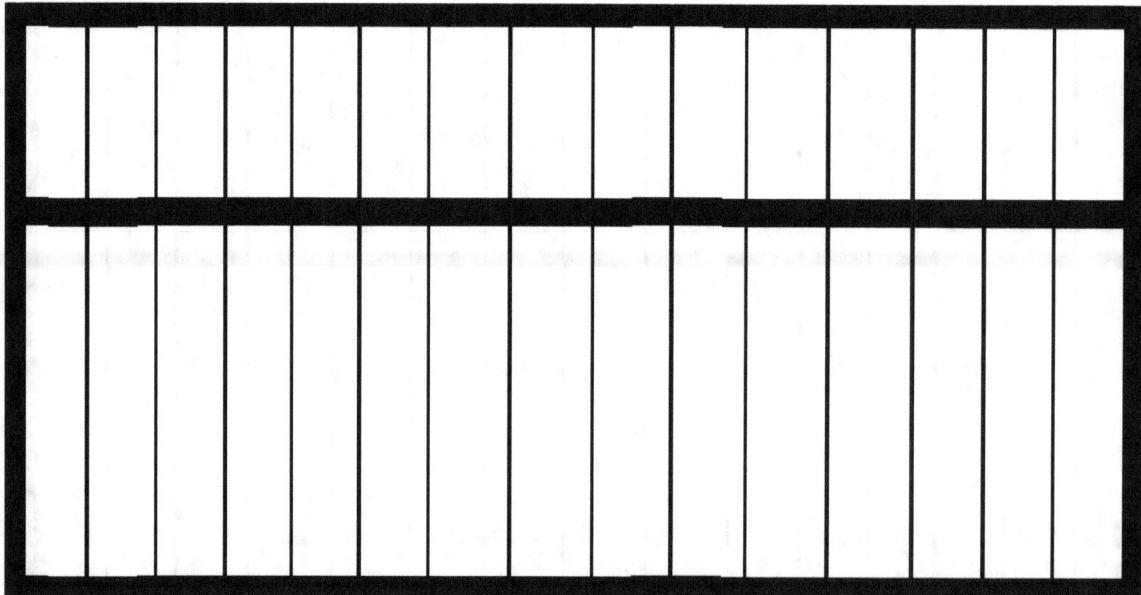

Test D, Show solution beads in final position.

Add a bar.

a. 16 x 7

b. 20 x 9

c. 13 x 4

d. (65 - 33) x 19

e. 12/4 x 64/8

f. 219 / 3

g. 33 x 47 +9 -18 / 7   *convert remainders to 2 decimal places*

h. (65 X 88) / 43

i. 23+19+37.9 + 3.5 - 12

j. $ .68 + $26.50 + $208 - $45.01

Complete all tests and then check the answers in this book, Subtract 2.5% for each wrong answer. An answer is wrong if any part of the answer is wrong. The answer may be placed anywhere on the Abacus, because you determine where the decimal point would be. There is no credit for completing more than what is asked for but you will give yourself an honorable mention! Add 2% for each wrong answer you are able to correct. If you score less than 75%, you need to start this book and all the work again.

100%   Abacus Master

80%   Abacus Expert

75%   Successful Compleation

Your Certificate is at the back of this book.

Let me know if you enjoyed learning how to use the Chinese Abacus.

Yours in Learning and Teaching, Joe Salazar

Jws345@hotmail.com

Intentionally left blank

Test A, Answers

a. 2   2   +   3   4   =   5   6

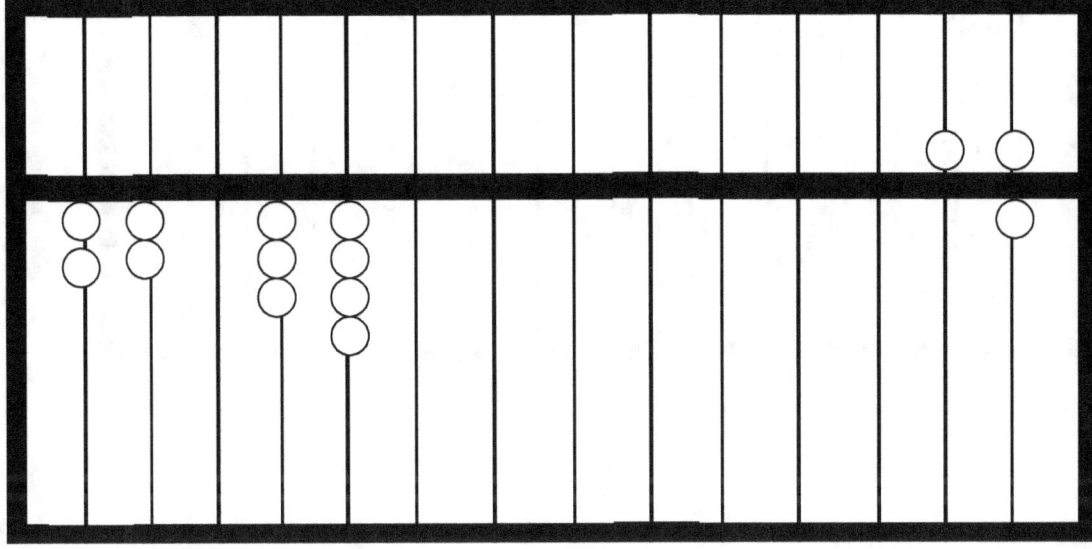

b. 1   6   +   2   6 =   4   2

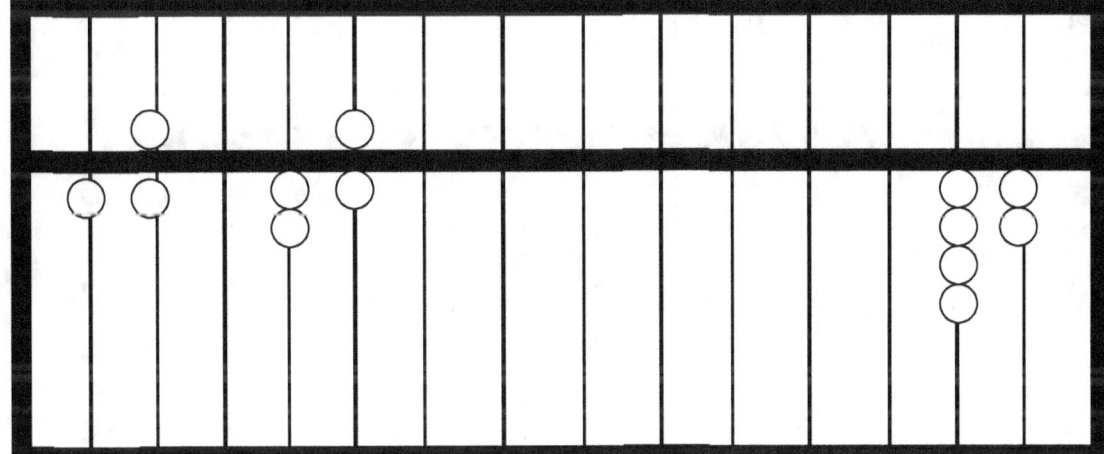

c. 104+220 =

3 2 4

d. 30+70 = 100

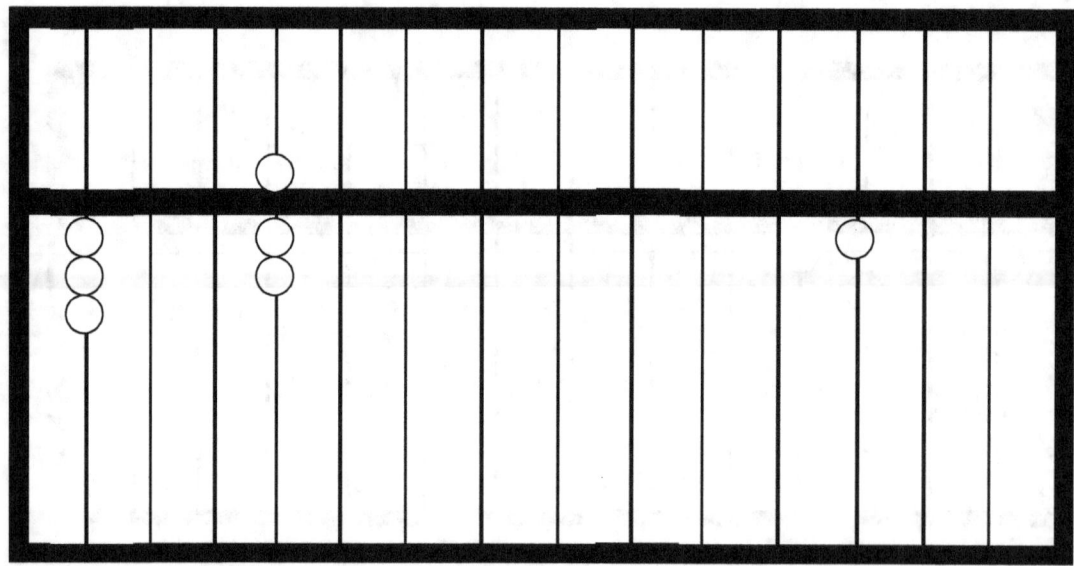

e. 4004+2030 = 6 0 3 4

f. 180 + 810 = 990

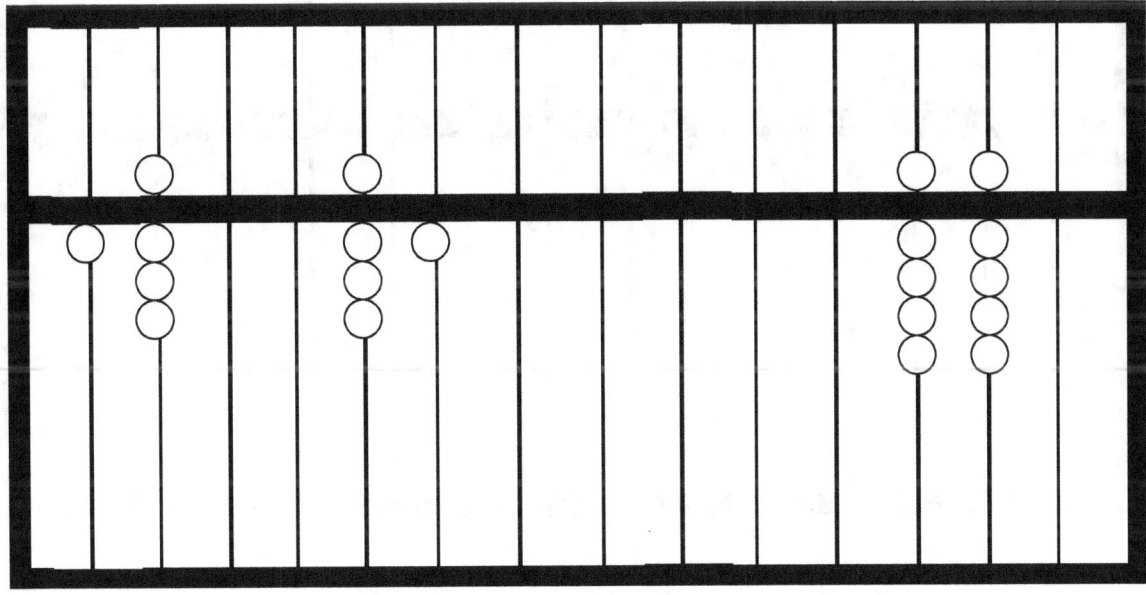

g. 999+499 = 1,498

h. 115+15 = 130

i. 203 + 219 = 422

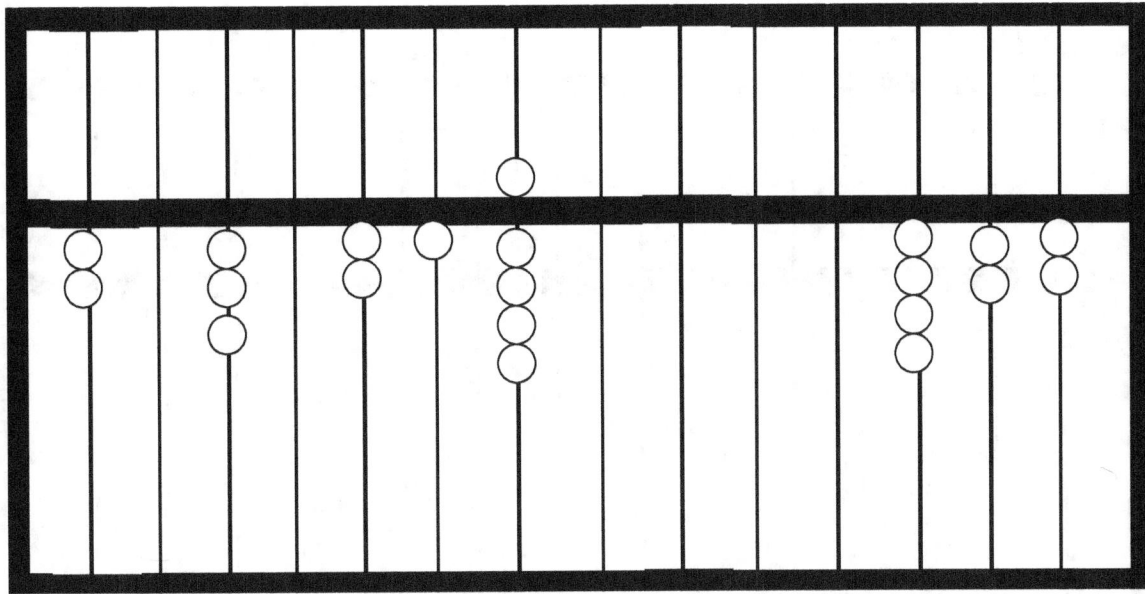

j. 4009 + 9090 = 13, 099

Test B, Answers

a. 35-18 = 17

b. 355-28 = 327

c. 298-120 = 178

d. 507-203 = 304

e. 3001-2003= 998

f. 100-97= 3

g. 222-199 = 23

h. 606-284 = 322

i. 903-605 = 298

j. 505-209 = 296

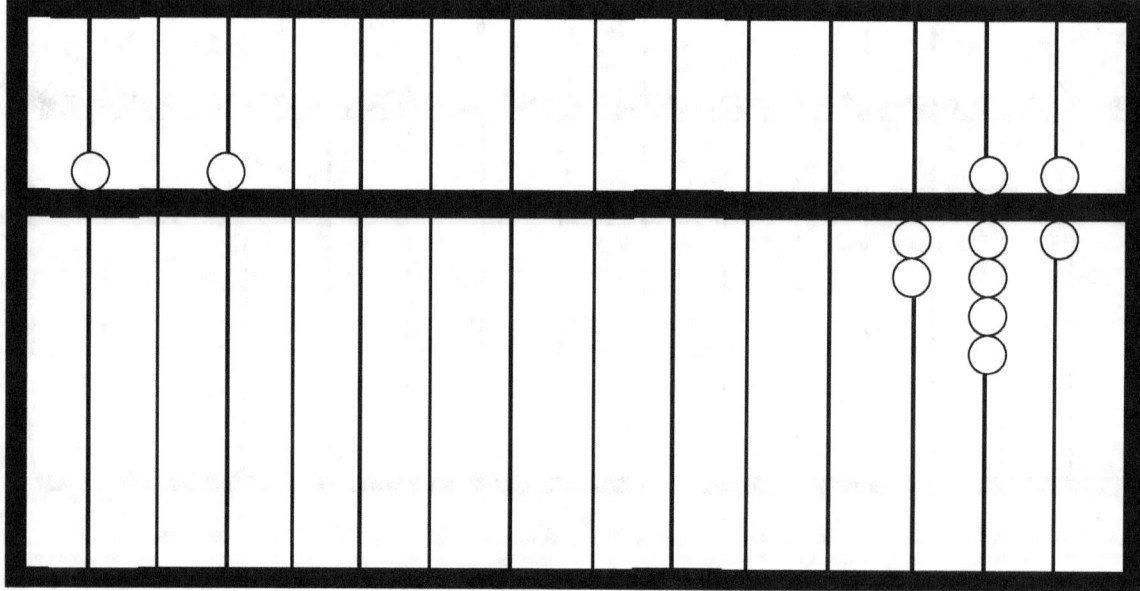

Test C, Diagram only the solution calculated on abacus

a. 23+18-9+17 =49

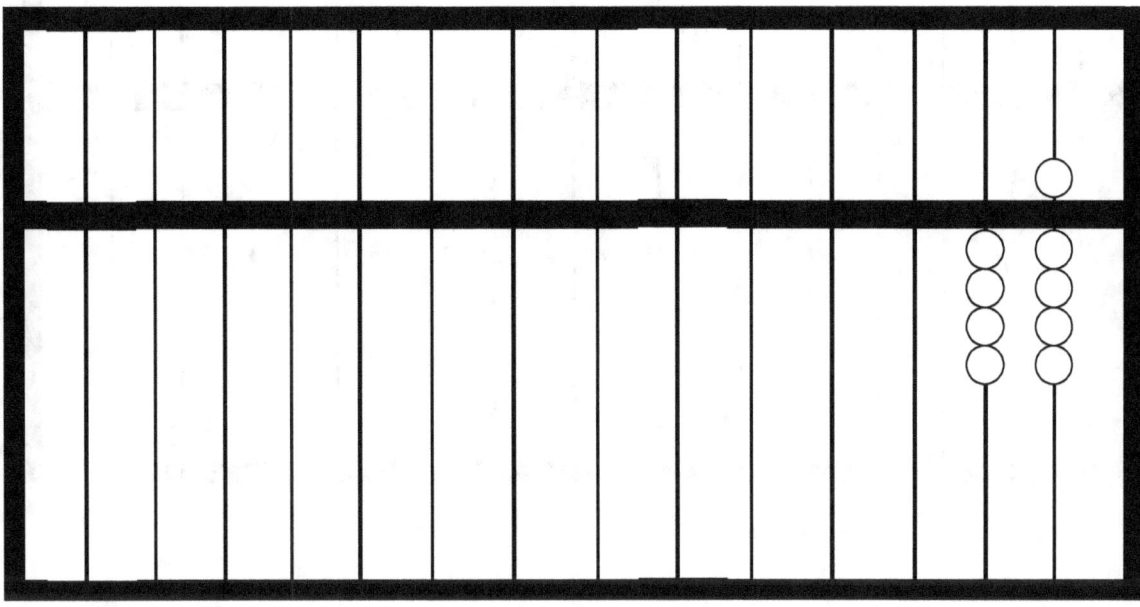

b. 304-210+190-12 = 272

c. 213-31+180 = 362

d. 108-9+19-26 = 92

e. 999-18+54-22+75 = 1088

f. 211+112-121+212-2 = 412

g. 309-27+91-100 = 273

h. 209-87+91-17+20 = 216

i. 120-19+109-15-22+390 = 563

j. 3400+602-1009 = 2993

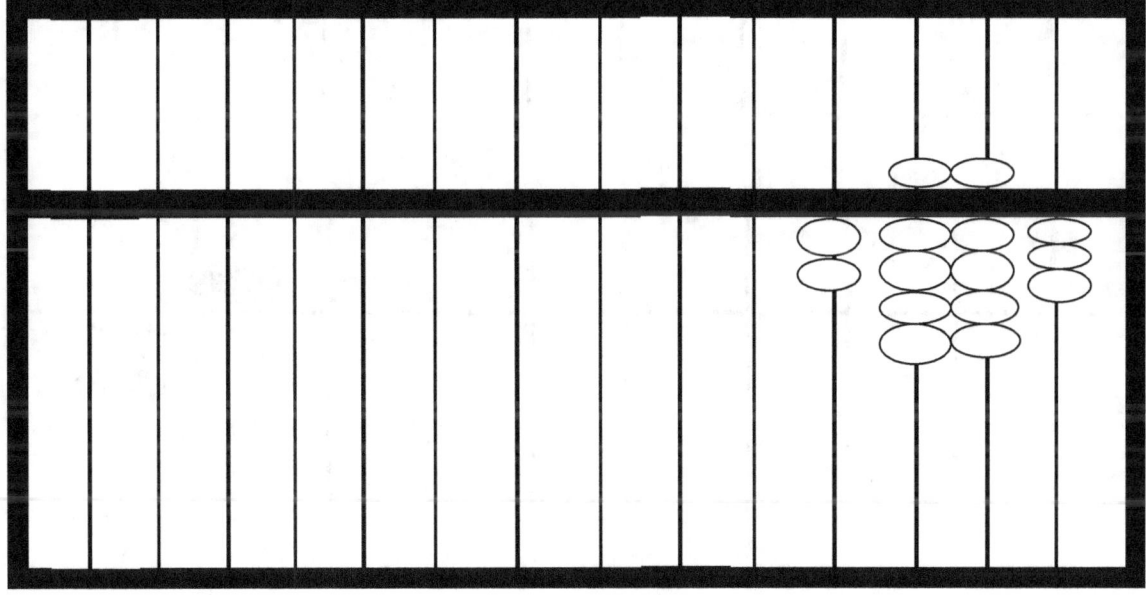

Test D, Answers

a. 16 x 7

b. 20 x 9

c. 13 x 4

d. (65 - 33) x 19

e. 12/4 x 64/8

f. 219 / 3

It is appropriate to plan a quotient from left to right because we don't know how many spaces we will need.

g. *33 x 47 +9 -18 / 7*   *convert remainders to 2 decimal places*

h. *(65 X 88) / 43*

i. *23+19+37.9 + 3.5 - 12*

Congratulations you have completed 10 hours, or more, of Abacus Training.

The Author

Joseph W. Salazar

Born Dec. 9, 1940 Chicago Illinois

Enlisted in the U.S. Army at age seventeen, April 1958

Assigned to the Medical Corps and served six years, four of which were in Germany. I worked in many types of Army Medical Clinics and hospitals as a Medical Attendant and Technician. After leaving the Army I lived in Salt Lake City for two years before moving to California and worked as a Vocational Nurse, Driving Instructor, Martial Arts Teacher, and Traffic Violator School Instructor, while obtaining my B. S. in Health Science at San Jose State University.

I got my J. D. from Saratoga University Law School, Distance Learning Program.

My hobbies are Chess, Martial Arts, Archery and Flying Planes.

I taught the Accounting Clerk Program for Goodwill Industries for three years and operated my own Tax & Accounting business for forty years. I continue to operate my own business as a Public Accountant.

Joe Salazar

Prunedale California, August 2023

Certificate of Achievement

Awarded to _____

for completion of training in the art of

Suanpan

Chinese Abacus

on this date _____

presented by instructor Joseph W. Salazar

www.ingramcontent.com/pod-product-compliance
Lightning Source LLC
Chambersburg PA
CBHW082144290526
45794CB00008B/3160